Pebble® Bilingüe/Bilingual Plus

¡Criaturas diminutas!/Bugs, Bugs, Bugs!

Saltamontes/Grasshoppers

por/by Margaret Hall

Traducción/Translation: Dr. Martín Luis Guzmán Ferrer
Editor Consultor/Consulting Editor: Dra. Gail Saunders-Smith

Consultor/Consultant: Gary A. Dunn, MS, Director of Education
Young Entomologists' Society Inc.

Capstone press®

Mankato, Minnesota

Pebble Plus is published by Capstone Press,
151 Good Counsel Drive, P.O. Box 669, Mankato, Minnesota 56002.
www.capstonepress.com

1 2 3 4 5 6 11 10 09 08 07 06

Library of Congress Cataloging-in-Publication Data
Hall, Margaret, 1947–
    [Grasshoppers. Spanish & English]
    Saltamontes = Grasshoppers/de/by Margaret Hall.
    p. cm.—(Pebble plus. ¡Criaturas diminutas!/Bugs, bugs, bugs!)
    Includes index.
    ISBN-13: 978-0-7368-6679-8 (hardcover)
    ISBN-10: 0-7368-6679-5 (hardcover)
    1. Grasshoppers—Juvenile literature. I. Title. II. Title: Grasshoppers. III. Series: Pebble plus. ¡Criaturas
diminutas! (Spanish & English)
QL508.A2H29 2007
595.7'26—dc22   2005037468

Summary: Simple text and photographs describe the physical characteristics and habits of grasshoppers—in
    both English and Spanish.

Editorial Credits
Sarah L. Schuette, editor; Katy Kudela, bilingual editor; Eida del Risco, Spanish copy editor; Linda Clavel,
    set designer; Kelly Garvin, photo researcher; Karen Hieb, product planning editor

Photo Credits
Bill Johnson, 6–7
Bruce Coleman Inc./Gary Meszaros, 4–5; Laura Riley, 20–21
David Liebman, 12–13, 16–17
Dwight R. Kuhn, 9, 19
Gerry Ellis & Michael Durham/PictureQuest, 1
Pete Carmichael, 11, 15
Robert & Linda Mitchell, cover

Note to Parents and Teachers

The ¡Criaturas diminutas!/Bugs, Bugs, Bugs! set supports national science standards related to the diversity of life and heredity. This book describes grasshoppers in both English and Spanish. The images support early readers in understanding the text. The repetition of words and phrases helps early readers learn new words. This book also introduces early readers to subject-specific vocabulary words, which are defined in the Glossary section. Early readers may need assistance to read some words and to use the Table of Contents, Glossary, Internet Sites, and Index sections of the book.

Table of Contents

Tabla de contenidos

Grasshoppers

What are grasshoppers?

Grasshoppers are insects.

Los saltamontes

¿Qué son los saltamontes?

Los saltamontes son insectos.

How Grasshoppers Look

Many grasshoppers have
green or brown bodies.
Grasshoppers also can
be other colors.

Cómo son los saltamontes

Muchos saltamontes tienen
los cuerpos verdes o marrones.
También hay saltamontes
de otros colores.

Grasshoppers are about
the size of a child's little
finger. Grasshoppers have
six legs and two wings.

Los saltamontes son como del
tamaño del dedo meñique
de un niño. Los saltamontes
tienen seis patas y dos alas.

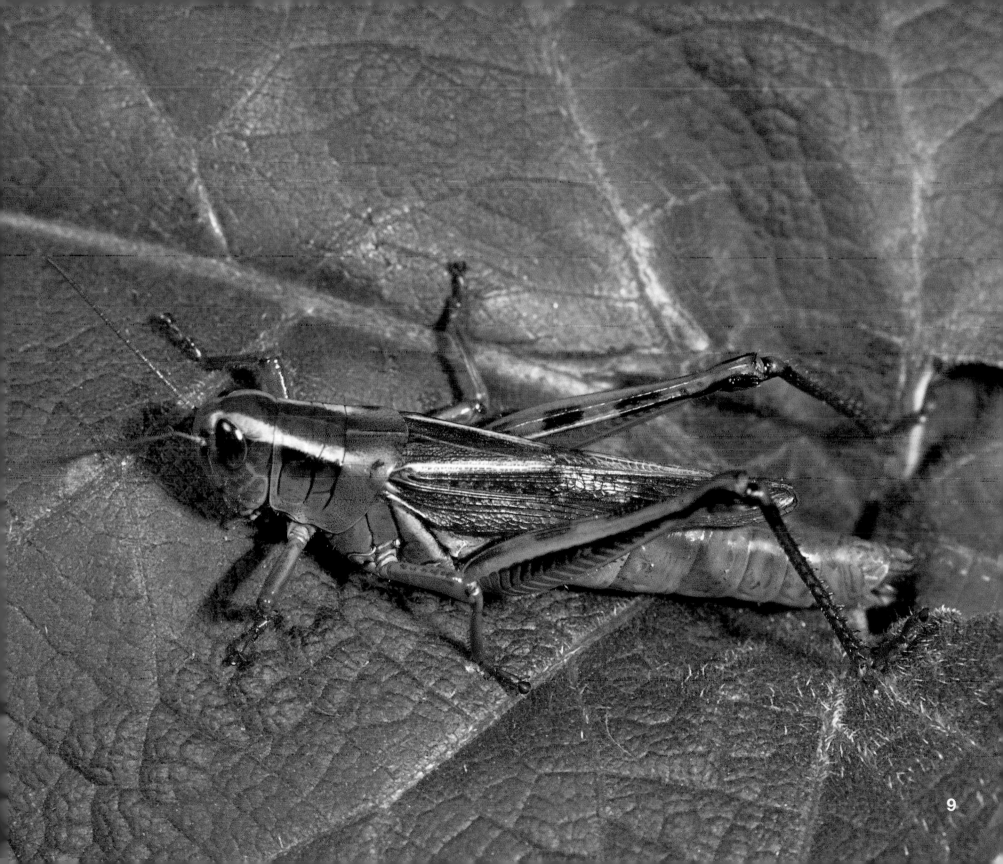

Grasshoppers have two
large eyes.

Los saltamontes tienen
dos ojos grandes.

Grasshoppers have two
antennas. Grasshoppers
feel and smell with
their antennas.

Los saltamontes tienen dos
antenas. Los saltamontes tocan
y huelen con sus antenas.

Grasshoppers have two
sharp jaws.

Los saltamontes tienen
dos mandíbulas filosas.

What Grasshoppers Do

Grasshoppers chew and

chomp on plants.

Qué hacen los saltamontes

Los saltamontes muerden y

mastican plantas.

Grasshoppers jump.
They use their long
back legs to jump high.

Los saltamontes saltan.
Usan sus largas patas
traseras para saltar
muy alto.

Male grasshoppers rub
their legs against their
wings to sing.

Los saltamontes macho
frotan sus patas contra
sus alas para cantar.

Glossary

antenna—a feeler; insects use antennas to sense movement, to smell, and to listen to each other.

insect—a small animal with a hard outer shell, six legs, three body sections, and two antennas; most insects have wings.

jaw—a part of the mouth used to grab, bite, and chew

male—an animal that can father young

Glosario

la **antena**—parte del cuerpo para tocar y sentir; los insectos usan las antenas para sentir el movimiento, olfatear y escucharse entre sí.

el **insecto**—animal pequeño con un caparazón duro, seis patas, cuerpo dividido en tres secciones y dos antenas; la mayoría de los insectos tiene alas.

el **macho**—que puede ser padre

la **mandíbula**—parte de la boca que se usa para atrapar, morder y masticar

Internet Sites

FactHound offers a safe, fun way to find Internet sites related to this book. All of the sites on FactHound have been researched by our staff.

Here's how:

1. Visit *www.facthound.com*

2. Choose your grade level.

3. Type in this book ID **0736866795** for age-appropriate sites. You may also browse subjects by clicking on letters, or by clicking on pictures and words.

4. Click on the **Fetch It** button.

FactHound will fetch the best sites for you!

Sitios de Internet

FactHound proporciona una manera divertida y segura de encontrar sitios de Internet relacionados con este libro. Nuestro personal ha investigado todos los sitios de FactHound. Es posible que los sitios no estén en español.

Se hace así:

1. Visita *www.facthound.com*

2. Elige tu grado escolar.

3. Introduce este código especial **0736866795** para ver sitios apropiados según tu edad, o usa una palabra relacionada con este libro para hacer una búsqueda general.

4. Haz clic en el botón **Fetch It**.

¡FactHound buscará los mejores sitios para ti!

Index

Índice